oxford **maths**

for australian schools

contents

OXFORD
UNIVERSITY PRESS

Practice

1 Complete the number lines.

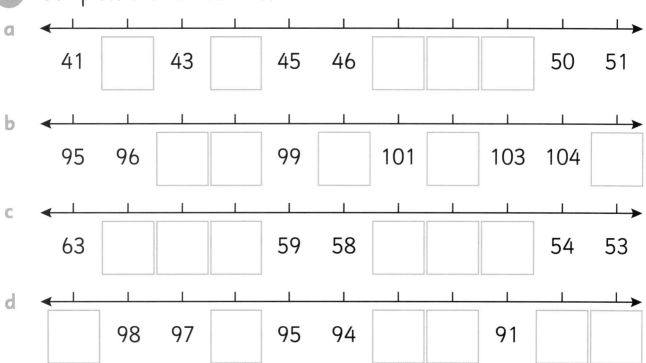

a
41 [ ] 43 [ ] 45 46 [ ] [ ] [ ] 50 51

b
95 96 [ ] [ ] 99 [ ] 101 [ ] 103 104 [ ]

c
63 [ ] [ ] [ ] 59 58 [ ] [ ] [ ] 54 53

d
[ ] 98 97 [ ] 95 94 [ ] [ ] 91 [ ] [ ]

2 Write these numbers in the correct places on the number chart.

19

82

104

45

67

33

119

6

90

71

1									
									120

Oxford University Press

1 Tom and Lexi play "First to 30". Help Lexi work out how many counters Tom has on his ten frames.

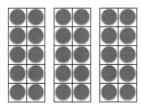

In "First to 30", the first person to make 30 by filling three ten frames wins!

a 6 spaces left = ☐ counters on the ten frames.

b 11 spaces left = ☐ counters on the ten frames.

c 18 spaces left = ☐ counters on the ten frames.

d 9 spaces left = ☐ counters on the ten frames.

e 23 spaces left = ☐ counters on the ten frames.

2 Ana and Stella live on the same street. The difference between their house numbers is 14. If there are 120 houses on their street, what might their house numbers be?

Ana's house number	Stella's house number

You might like to use a number line to help you.

1 Lexi rolls a dice six times and makes exactly 30. What numbers could she have rolled?

a _____ + _____ + _____ + _____ + _____ + _____ = 30

b _____ + _____ + _____ + _____ + _____ + _____ = 30

c _____ + _____ + _____ + _____ + _____ + _____ = 30

d _____ + _____ + _____ + _____ + _____ + _____ = 30

e _____ + _____ + _____ + _____ + _____ + _____ = 30

2 Guess the numbers.

a I am an even number. The difference between the two digits in my number is 2. What number might I be?

b I am an odd number. The difference between my two digits is 3. What number might I be?

Oxford University Press

Practice

1 Write these house numbers in words.

17 _____

41 _____

66 _____

103 _____

2 Match each person's age to the correct birthday cake.

27

12

50

18

35

twenty-seven

thirty-five

fifty

1 Roll two dice and use the numbers to make a 2-digit number. Write the number in numerals and words. Draw it in pictures, e.g. cubes.

Number (numerals)	Number (words)	Number (pictures)

2 What other 2-digit numbers could you have made from the numbers you rolled in question 1?

My number	Other number I could have made

I rolled a 3 and a 5. I could have made 35 or 53!

3 What number is shown by **x** on the number lines? Write it in numerals and words.

Number line	Number (numerals)	Number (words)
18 19 20 X 22 23 24		
106 107 108 109 X 111 112		
27 X 25 24 23 22 21		
43 42 X 40 39 38 37		
89 X 91 92 93 94 95		

Oxford University Press

1 Jess saw beetles crawling up a tree. Write how many beetles she might have seen. Each beetle has six legs. Work out how many legs.

Number of beetles	Number of legs (numerals)	Number of legs (words)
e.g. 2	12	twelve

2 There are seven 2-digit numbers whose digits when added together equal 7. Write each of them in numerals and in words.

e.g. 16   sixteen

1 and 6 is 7. What other numbers can be combined to make 7?

Practice

1 Put these numbers in the correct places on the number lines.

a 54   b 39   c 43

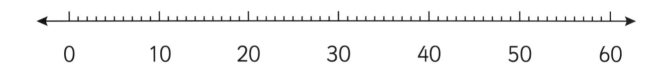

d 66   e 51   f 79

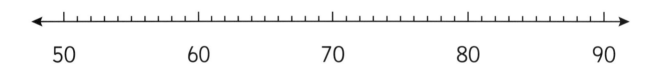

2

a Roll two 10-sided (0–9) dice. Combine the numbers to make a 2-digit number. Do this six times. Record the numbers below.

_____ _____ _____ _____ _____ _____

b Order your six numbers from smallest to largest.

Smallest   Largest

_____ _____ _____ _____ _____ _____

Oxford University Press

1 Roll two 10-sided dice. Combine the numbers to make eight 2-digit numbers. Write them below.

_____ _____ _____ _____ _____ _____ _____ _____

Use your numbers to answer the following questions.

a Which number is the largest? _____

b Which number is the smallest? _____

c Which numbers are odd? _____

d Which numbers are even? _____

e Which numbers are less than 50? _____

f Which numbers are more than 50? _____

2

a How many different 2-digit numbers can you make from these numbers?

2   7   4   5

b Order the numbers you made from largest to smallest.

_____

_____

1 This is the basketball ladder for the top 7 teams in the league.

Team	Red	Blue	Green	Brown	White	Pink	Black
Scores	24	22	20	18	18	16	15

a Below are the scores for the latest games. Will these scores change the ladder? _____

Team	Red	Blue	Green	Brown	White	Pink	Black
Scores	5	12	7	2	10	6	8

b Complete the new ladder.

Team							
Scores							

2 Roll two 10-sided dice to make a 2-digit number. Write it under "My number" in the table. Write the numbers that come before and after your number.

Number before	My number	Number after

3 Try it with 3-digit numbers. Roll three dice or ask someone to tell you a 3-digit number.

Number before	My number	Number after

Oxford University Press

Practice

1. Miss Wu asks Leo to solve 6 + 13. Leo starts from the number 13. Why? Explain your thinking and show your working out.

2. Use the same strategy to solve these calculations. Show your working out.

   a 9 more than 11

   b 7 and 12

3. Fill the gaps with the numbers to make a correct number sentence. Show each one on the number line using counting on.

   a 16   20   4

   b 17   9   8

   _____ + _____ = _____

   _____ + _____ = _____

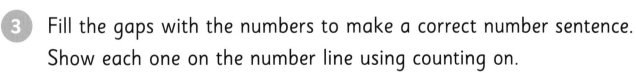

1 Max has 6 toy cars. For his birthday, he was given more as gifts. How many cars might Max have now?

6 + ☐ = ☐

6 + ☐ = ☐

6 + ☐ = ☐

6 + ☐ = ☐

6 + ☐ = ☐

2 Roll two 6-sided dice to make a 2-digit number and add them together to get your start number. Record your start number. Count on 8 more and write the new number.

☐ + 8 = ☐

☐ + 8 = ☐

☐ + 8 = ☐

☐ + 8 = ☐

☐ + 8 = ☐

3 Myra's friend gave her some shells. She has 17 shells now. How many might Myra have started with? How many did her friend give her?

☐ + ☐ = 17

☐ + ☐ = 17

☐ + ☐ = 17

Oxford University Press

1 Julie and Maria play a game. Julie wins by 11 points. What might their scores be?

Julie's score	Maria's score

You might like to use a hundred chart to help you.

2 Bill's birthday is in June. Paul's birthday is 14 days after Bill's. When might each of their birthdays be?

June

				1	2	3
4	5	6	7	8	9	10
11	12	13	14	15	16	17
18	19	20	21	22	23	24
25	26	27	28	29	30	

July

						1
2	3	4	5	6	7	8
9	10	11	12	13	14	15
16	17	18	19	20	21	22
23	24	25	26	27	28	29
30	31					

Bill's birthday	Paul's birthday

Practice

1 Partition the numbers.

a

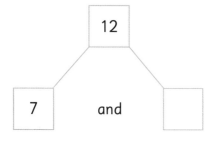

12

7 and ☐

b

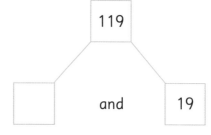

119

☐ and 19

c

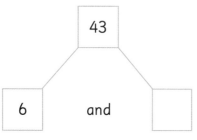

43

6 and ☐

d

38

☐ and 16

2 The pictures on the left show the same amounts as the pictures on the right. Find the matching pairs.

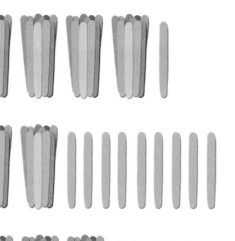

T O

T O

T O

T O

Challenge

1 Find three different ways to partition each number.

Number	Partition 1	Partition 2	Partition 3
13			
18			
21			
35			
62			
107			

2 Amelia had lots of teddy bears. She gave half of them away. How many bears might Amelia have started with? How many did she give away?

Started with	Gave away

1 How many different ways can you partition the number 110?

2 Jack gives nine gel pens to Ava.
How many pens might Jack have started with?
How many does he have now?

You can use this diagram to help you.

9 and

Oxford University Press

Practice

1 Mr Kumar writes $18 - 7 =$ ☐ on the board.

a Use the number line to solve the problem.

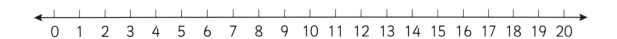

b Now solve these problems.

$16 - 7 =$ ☐ $14 - 9 =$ ☐ $19 - 11 =$ ☐

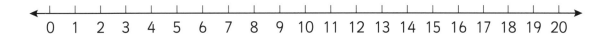

---

2 Bob and Peter play "First to 20". Bob wins all five games! Look at how many squares

In "First to 20", each player rolls a dice to make a number. They place that number of counters onto their ten frames. The first to make 20 by filling two ten frames wins!

Peter has left uncovered. Work out Peter's score for each of the five games.

	Squares uncovered	Peter's score (squares covered)
Game 1	3	
Game 2	6	
Game 3	11	
Game 4	14	
Game 5	9	

1 Roll two 10-sided dice. Combine the digits to create a number. Record your number. Count back 7 from your number and write the answer.

Your number   Answer   Your number   Answer

[ ] – 7 = [ ]   [ ] – 7 = [ ]

[ ] – 7 = [ ]   [ ] – 7 = [ ]

[ ] – 7 = [ ]   [ ] – 7 = [ ]

[ ] – 7 = [ ]   [ ] – 7 = [ ]

[ ] – 7 = [ ]   [ ] – 7 = [ ]

2 Roll a 10-sided dice. Record the number. Count back that many from 26 and write the answer.

Number rolled   Answer   Number rolled   Answer

26 – [ ] = [ ]   26 – [ ] = [ ]

26 – [ ] = [ ]   26 – [ ] = [ ]

26 – [ ] = [ ]   26 – [ ] = [ ]

26 – [ ] = [ ]   26 – [ ] = [ ]

26 – [ ] = [ ]   26 – [ ] = [ ]

Could you use a number line to help you?

1 Lucas had 32 marbles but he dropped them and lost some. How many marbles might Lucas have lost? How many are left?

2 Alice missed out on 9 points on her maths test. Decide what the maximum possible test score could be. Use the counting back strategy to work out Alice's score. Repeat with different maximum scores.

Maximum possible test score	Alice's test score
e.g. 40	31

Practice

1 Find the difference between the two numbers.

You could use blocks to help you, or draw a number line.

a   15 and 7 ☐   b   19 and 13 ☐

c   4 and 12 ☐   d   11 and 23 ☐

e   25 and 9 ☐   f   13 and 29 ☐

2 Choose two numbers between 0 and 100. Use the number line to find the difference between your chosen numbers.

My two numbers are _____ and _____ .

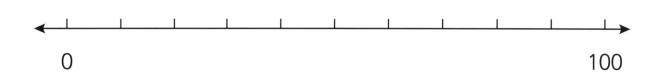

0   100

Oxford University Press

1 Zoe is thinking of two numbers. The difference between her two numbers is 7. What could Zoe's two numbers be?

☐ and ☐   ☐ and ☐

☐ and ☐   ☐ and ☐

☐ and ☐   ☐ and ☐

☐ and ☐   ☐ and ☐

2 Roll two 10-sided dice to make two 2-digit numbers. Find the difference between the first and second numbers.

First number	Second number	Difference
e.g. 56	65	The difference is 9.

You could use a number line to help you.

+9

50   56 60   65 70

1 The red team played the blue team at football. The red team scored more than double the blue team's score. What might each team have scored? Find the difference between their scores.

Red team's score	Blue team's score	Difference

2 Sofia put 31 candles on her mum's birthday cake. Dad says Mum is going to be older than that. Decide how old Mum might be turning and how many more candles Sofia should add.

How old will Mum be?	How many more candles?
e.g. 35	4 more candles

Oxford University Press

Practice

1 Grouping items into groups of five can make counting larger collections easier. Count by 5s to find the amounts below.

a

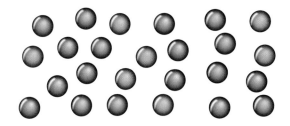

Amount: _____

b

Amount: _____

c

Amount: _____

2 Skip counting by 2s can also be a quick way to count. How many eyes are on the people below?

Eyes: _____

Eyes: _____

Eyes: _____

1 Skip count to complete the blanks.

a   0 [ ] [ ] [ ] 20 [ ] 30

b   [ ] 15 [ ] [ ] 30 [ ] [ ]

c   18 [ ] [ ] [ ] 28 [ ]

d   [ ] [ ] 45 [ ] [ ] 30 [ ]

e   [ ] 50 [ ] [ ] [ ] [ ] 0

2 Make skip counting patterns. Start from a number of your choice.

a   Skip count by 2s.

b   Skip count by 5s.

c   Skip count by 10s.

3 Make two patterns that skip count **backwards**. Decide which numbers to start from.

a   Skip count by 2s.

b   Skip count by 5s.

Oxford University Press

1 Mia creates a skip counting pattern that lands on 40. What might Mia's skip counting pattern look like?

Skip count by	Skip counting pattern

2

a A group of people are at the bus stop. How many eyes are there altogether?

Number of people	Skip counting pattern	Total eyes

b A florist sells bunches of flowers with 10 flowers in each bunch. How many flowers are there altogether?

Number of bunches	Skip counting pattern	Total flowers

Practice

1 Work out the answers.

a   10 shared between 2.

b   12 shared between 3.

c   15 shared between 5.

d   16 shared between 4.

---

2 Oliver gives his friends 4 biscuits each. How many biscuits might Oliver have to share?

How many friends?	Equal shares (draw 4 biscuits each)	Total biscuits

Oxford University Press

1 Isla has 12 lollies to share. How many will each person get if she shares the lollies between:

a two people?

b three people?

c four people?

d six people?

2

a Can Nico share nine lollies between three people? Show your working out.

> When you are sharing equally, make sure everyone gets the same amount.

b Can Nico share nine lollies equally between four people? Show your working out.

1 How many people could Isabelle share 20 lollies equally between?

2 Zac gives each friend three lollies. How many lollies might Zac have?

How many friends?	Equal groups (draw)	How many lollies in total?

Practice

1 Circle all the pictures with equal halves.

2 Circle all the pictures with equal quarters.

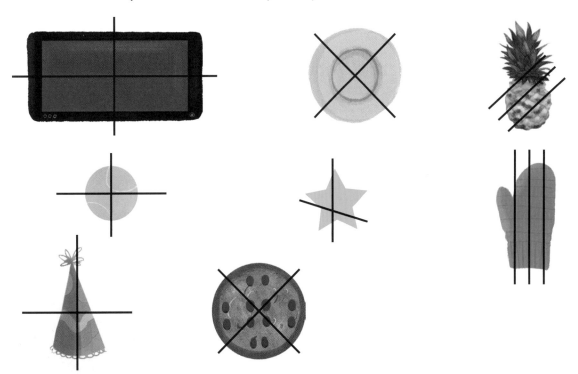

1 Show four different ways the squares could be cut into quarters.

2

a Molly ate $\frac{1}{2}$ of her pizza. What might Molly's pizza look like now?

b Nico ate $\frac{1}{4}$ of his pizza. What might Nico's pizza look like now?

c Frank ate $\frac{3}{4}$ of his pizza. What might Frank's pizza look like now?

If you have eaten $\frac{3}{4}$ of your pizza, how much is left?

Oxford University Press

1 Design a mat that is $\frac{1}{4}$ red, $\frac{1}{2}$ blue and $\frac{1}{4}$ green.

2 Create your own pizza. What toppings will you use? You can have different toppings on different parts of the pizza.

Fraction   **Topping**

_____ of my pizza has _____ .

_____ of my pizza has _____ .

_____ of my pizza has _____ .

_____ of my pizza has _____ .

Practice

1. Circle half of each group.

a 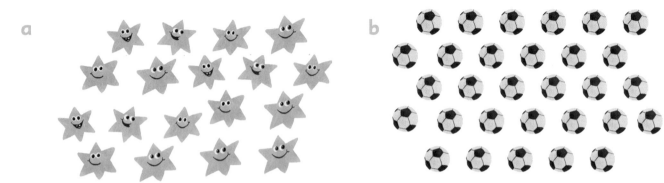 b

2. Circle one quarter of each group.

a  b

3. Draw circles to divide the group into halves.

a How many groups? ☐   b How many in each group? ☐

Oxford University Press

How many different ways can you find to show this?

1 Sam and his mum made cookies. Half the cookies were plain and half were chocolate chip. Draw what the cookies might look like.

2 Timmy and Kaden are playing marbles. One quarter of their marbles are blue. Draw what their marbles might look like below. How many different ways can you show this?

1 Work out the answers.

Question	Working out
There are 24 students in Sienna's class. Half are boys. How many boys are there?	
Dan has 20 people in his class. $\frac{1}{4}$ of them are away today. How many are away?	
There are 30 people watching rugby. Half are wearing jumpers. How many people are wearing jumpers?	
Jess has 16 people in her class. $\frac{1}{4}$ are away today. How many of her class are at school?	

2 Rose pulled a handful of counters out of a container. $\frac{1}{2}$ were red, $\frac{1}{4}$ were green and the rest were yellow. How many of each colour might there be?

Total counters	Red counters	Green counters	Yellow counters

Practice

1

a   How many different Australian coins are there? Draw them all below.

b   How many coins are gold? _____

c   How many are silver? _____

d   Which coin is the biggest? _____

e   Which is the smallest? _____

f   Which coin is worth the most? _____

g   Which is worth the least? _____

2   Match the coins that add up to the same amount.

1 Mrs Liang asks Class 1 to make $1 using only silver coins.
Show three different ways they could make $1.

2 Circle the larger amount.

a

 or

b

    or

c

   or

d

  or

1 I have one gold and one silver coin. How much money might I have? Draw some possible answers.

2 I have $2.65 in my purse. How many coins might I have? Draw what they might look like.

How many coins?	Draw the coins

Practice

1

a Put these coins in order of size from smallest to largest.

Smallest   Largest

b Now put them in order from least value to the most value.

Least   Most

2 Using the information above, answer these questions.

a Which coins are larger than the $1? _____

b Which coins are smaller than the 20c? _____

c Which coins are worth more than the 50c? _____

d Which coins are worth less than the 50c? _____

e Which coins can you double to equal the value of another coin?

_____

_____

1 What coins would you use to buy these items? Draw the coins in order from most to least value.

Item	Draw the coins (most value to least value)
$1.50	
$3.00	
$2.25	
$3.95	
$5.10	

2 Order the coins from most to least value. Add them to find the total.

Coins	Order	Total
20 10 50 5		
5 2 DOLLARS 5 1 DOLLAR 50		
20 1 DOLLAR 10 20 50		
10 1 DOLLAR 5 20		
2 DOLLARS 1 DOLLAR 5 50 20		

1

a Sometimes you can use more than one coin to make the same value as a single coin. How many coins can you do this with?

b Can you find examples where more than two coins are equal in value to a single coin?

I can use five 10-cent pieces to make 50 cents!

2 Ben found three coins under his bed. Two were the same and one was different. What might the coins be and how much might Ben have?

Coins	Amount

Oxford University Press

Practice

1 Complete the patterns.

a

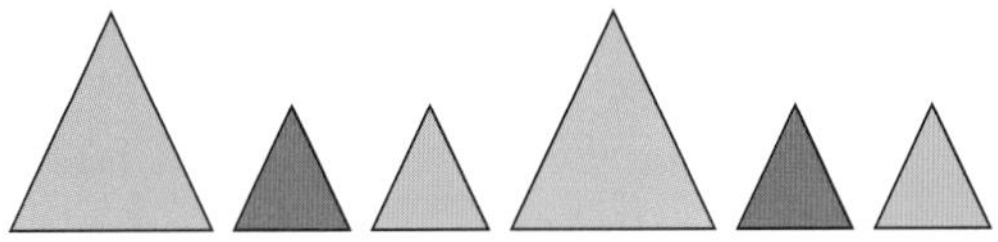

b

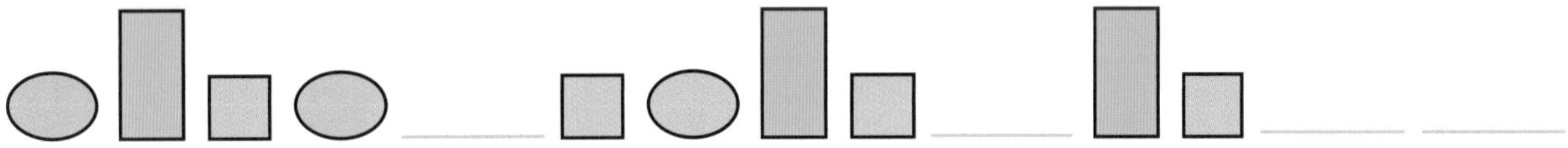

c

M W A M W A

d

P P R P P R ___ P R P P ___

---

2 Circle the error in each pattern. Complete the table.

Circle the error in each pattern	What should it be?	Rule for the pattern
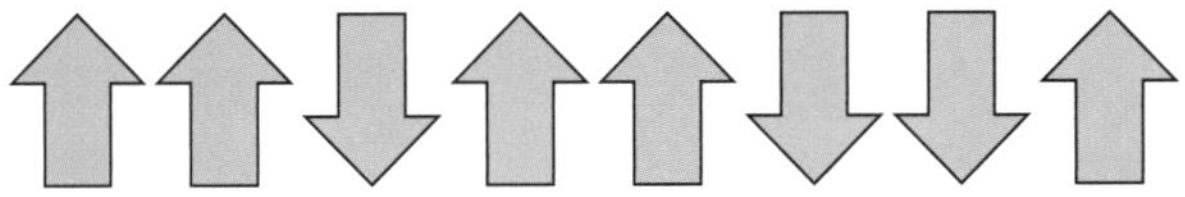		
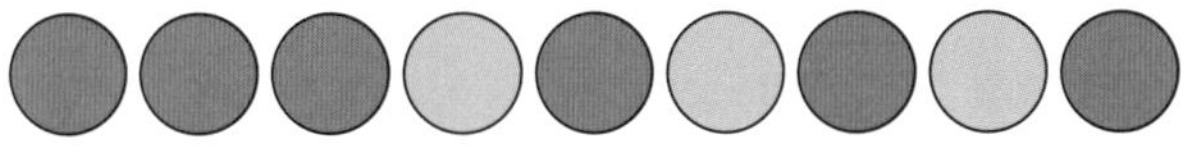		
X Y Z X Y Z X X Z		
O U U O O U O O U		
I I J I I I J J I I J J		

1 Create your own patterns. What is the rule for each pattern?

Rule: _____

Rule: _____

2 These patterns grow each time. Draw the next one in each pattern.

a

b

c

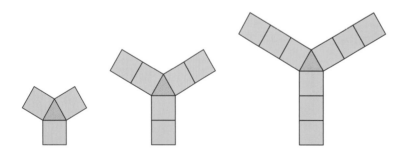

Oxford University Press

1 Izzy's garden path has stepping stones that are in a pattern. What might the pattern look like?

I wonder if it is a shape pattern, a colour pattern or a size pattern?

2 Ben opened a packet of lollies. He grouped the different coloured lollies together. He noticed that they made a growing pattern. What pattern might the lollies make?

Practice

1 Finish the number patterns.

a 36, 38, 40, 42, 44, _____ _____ _____ _____

b 55, 60, 65, 70, 75, _____ _____ _____ _____

c 120, 110, 100, 90, 80, _____ _____ _____ _____

d 80, 78, 76, 74, 72, _____ _____ _____ _____

e What does each pattern go up or down by?

Pattern a: _____   Pattern b: _____

Pattern c: _____   Pattern d: _____

2 Fill in the missing numbers to complete the patterns.

a 60, 70, ☐, 90, ☐, ☐, 120, 130, ☐

b 110, 105, 100, ☐, ☐, 85, ☐, ☐, 70

c 92, ☐, 96, 98, ☐, ☐, ☐, 106, 108

d 190, ☐, ☐, 160, ☐, ☐, 130, 120

e What does each pattern go up or down by?

Pattern a: _____   Pattern b: _____

Pattern c: _____   Pattern d: _____

Oxford University Press

Challenge

1

a Colour in the hundred chart to show a pattern.

b What is your pattern?

_____

c What number comes next in your pattern?

_____

1	2	3	4	5	6	7	8	9	10
11	12	13	14	15	16	17	18	19	20
21	22	23	24	25	26	27	28	29	30
31	32	33	34	35	36	37	38	39	40
41	42	43	44	45	46	47	48	49	50
51	52	53	54	55	56	57	58	59	60
61	62	63	64	65	66	67	68	69	70
71	72	73	74	75	76	77	78	79	80
81	82	83	84	85	86	87	88	89	90
91	92	93	94	95	96	97	98	99	100

2

a Colour a pattern that counts in 10s but does **not** start at 10.

1	2	3	4	5	6	7	8	9	10
11	12	13	14	15	16	17	18	19	20
21	22	23	24	25	26	27	28	29	30
31	32	33	34	35	36	37	38	39	40
41	42	43	44	45	46	47	48	49	50
51	52	53	54	55	56	57	58	59	60
61	62	63	64	65	66	67	68	69	70
71	72	73	74	75	76	77	78	79	80
81	82	83	84	85	86	87	88	89	90
91	92	93	94	95	96	97	98	99	100

b Colour a pattern that counts in 5s but does **not** start at 5.

1	2	3	4	5	6	7	8	9	10
11	12	13	14	15	16	17	18	19	20
21	22	23	24	25	26	27	28	29	30
31	32	33	34	35	36	37	38	39	40
41	42	43	44	45	46	47	48	49	50
51	52	53	54	55	56	57	58	59	60
61	62	63	64	65	66	67	68	69	70
71	72	73	74	75	76	77	78	79	80
81	82	83	84	85	86	87	88	89	90
91	92	93	94	95	96	97	98	99	100

1

a One of the numbers in Mo's pattern is 15. What might Mo's pattern look like?

b One of the numbers in Kim's pattern is 28. What might Kim's pattern look like?

2 House numbers in a street often go up in 2s. The numbers on Jack's street make a different pattern. What might the house numbers on Jack's street look like?

My house is number 18. Do you know what number your house is?

Practice

1 Here are five different-sized items.

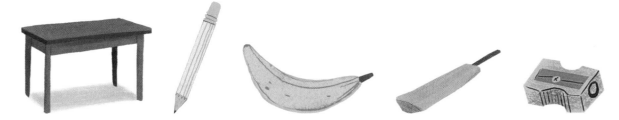

Order the items from shortest to longest in real life.

2 Find the area of the following shapes.

a _____ tiles

b _____ tiles

c _____ tiles

d _____ tiles

Challenge

1 Use paperclips to estimate and measure the length of these items.

Item	Estimate	Measure

2 Use blocks to estimate and measure the area of these items.

Item		Estimate	Measure
A box lid			
A poster			
A wall			

How close were your estimates? _____

Oxford University Press

1 Find items that are shorter than, the same length as, or longer than this book.

Shorter than	Same length as	Longer than

2 How many different shapes can you create with an area of nine squares?

Practice

1. Find the volume of these objects.

   a _____ cubes   b _____ cubes

   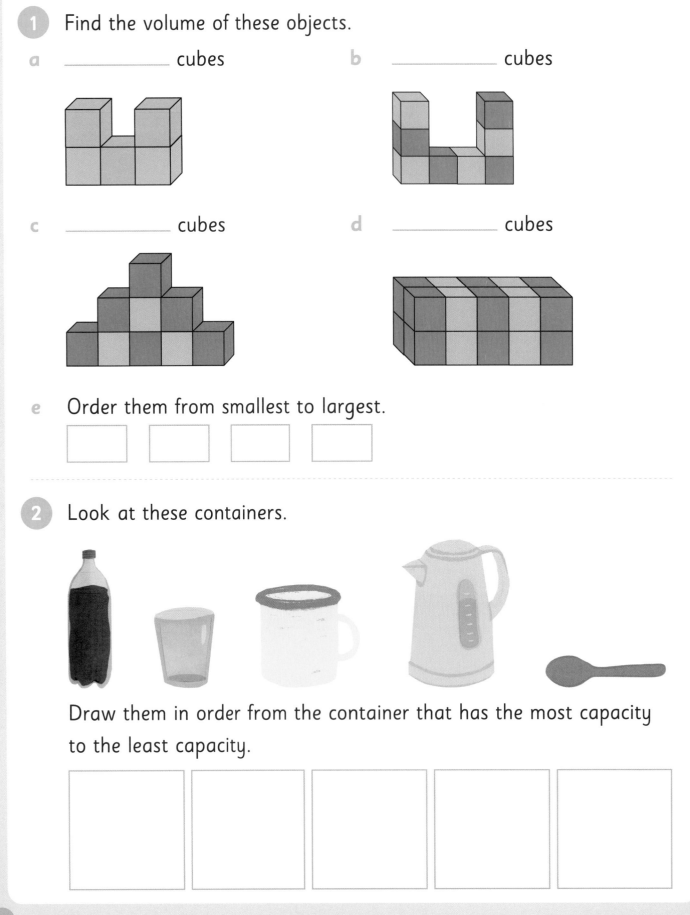

   c _____ cubes   d _____ cubes

   e Order them from smallest to largest.

2. Look at these containers.

   Draw them in order from the container that has the most capacity to the least capacity.

Challenge

1 Match the picture with its correct volume in cubes.

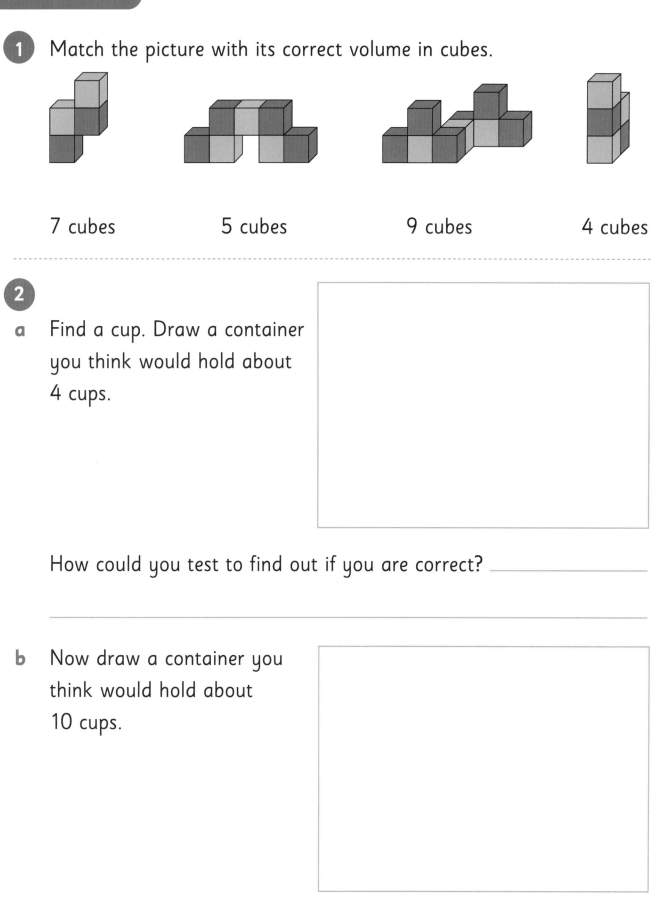

7 cubes   5 cubes   9 cubes   4 cubes

2

a Find a cup. Draw a container you think would hold about 4 cups.

How could you test to find out if you are correct? _____

b Now draw a container you think would hold about 10 cups.

How could you test to find out if you are correct? _____

1 Draw different objects, each with a volume of 8 cubes.

2

a Draw some containers with a larger capacity than a coffee cup.

b Draw some containers with a smaller capacity than a coffee cup.

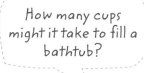
How many cups might it take to fill a bathtub?

Oxford University Press

Practice

1 Circle the heavier item in each pair.

a  or

b  or

c  or

d  or

2 Find five items that are heavier and five items that are lighter than this maths book.

Heavier	Lighter

1

a Draw five different items you can see around you.

b Order your five items from lightest to heaviest.

2 Draw pairs of items that you think weigh about the same.

a

b

c

You might need to do some hefting!

1 Draw or write an item on each side of the balance scales to make these true.

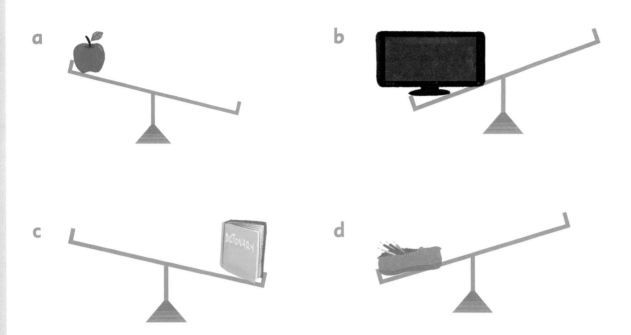

a

b

c

d

- -

2 One glue stick weighs the same as 4 cubes and one eraser weighs 3 cubes. Use this information and the pictures to answer the questions.

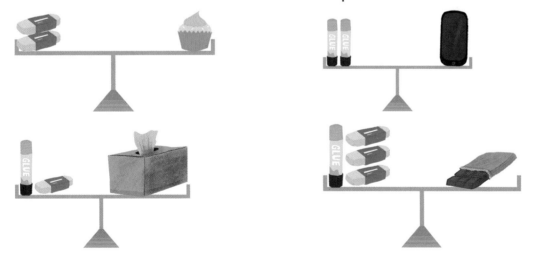

a How many cubes does the cake weigh? _____ cubes

b How many cubes does the phone weigh? _____ cubes

c How many cubes do the tissues weigh? _____ cubes

d How many cubes does the chocolate weigh? _____ cubes

Practice

1 Match the analog clock time to the digital clock time.

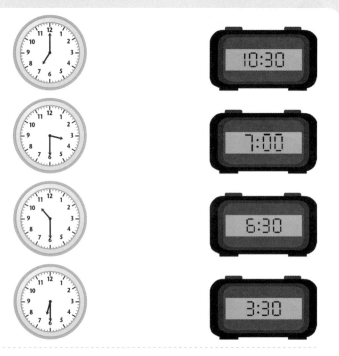

2 Write the time in numbers, words and draw the clock hands.

Digital time	The time in words	Analog time
11:30		
:		
:	nine o'clock	
:		
:	half past one	

Oxford University Press

1 What times are these clocks showing?

a  _____

b  _____

c  _____

d  _____

e  _____

f  _____

- -

2 Draw the times on the clocks.

a 10:30

b 5 o'clock

c 3:30

d 12:30

e 6 o'clock

f 7:30

1

a   Show your favourite time of the day on the two clocks.

b   What would you be doing at this time of the day?

_____

_____

2 One of the hands has fallen off the clock. The hand that is left is pointing at the 6. What might the time be?

   Oxford University Press

Practice

1 Put these events in order from shortest to longest duration.

a   Have a drink

b   Watch a movie

c   A day at school

d   Eat dinner

2 Would you measure the time until these events in hours, days, weeks, months or years?

Event	Time measured in
Your next birthday	
You finish primary school	
The weekend	
Winter	

1 Match the event with the time you would measure it in.

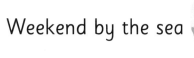

Weekend by the sea    Hours

The school holidays    Minutes

A school day    Weeks

Play on a seesaw    Days

2 Draw or write events on the timeline to show what you would do at each time on a special day with your family.

8 in the morning   11 in the morning   3 in the afternoon   7 in the evening

1 Order the birthdays of the people in your family.

Don't forget to include your own birthday!

2 Draw or write some things you could do in:

a one hour.

b one minute.

Practice

1 Here are some 2D shapes.

a Sort the shapes into groups.

b How did you sort the shapes? Give each group a heading.

_____

c How many different groups did you have? _____

2 Write two things that are the same and one thing that is different about each pair of shapes.

a

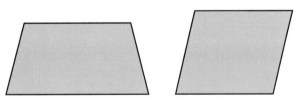

b

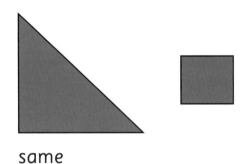

same

_____

_____

different

_____

same

_____

_____

different

_____

Challenge

1 Guess my shape.

a I have _____ corners. I have two short and two long sides.

What am I? _____

b Write your own "guess my shape" question for someone to solve.

```



```

What am I? _____

2 Draw some 2D shapes and describe their features.

Shape	Description of features

1 Draw a picture using only shapes. Use as many different shapes as you can!

2 Label the shapes in your picture. Record below how many of each shape you used.

Shape	How many in the picture?

Practice

1 Draw lines to match each 3D object to its name.

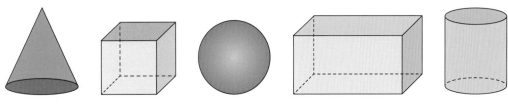

| cube | rectangular prism | cone | sphere | cylinder |

2 Draw and name the objects that match these descriptions.

Description	Draw and name the object
6 faces	
0 edges	
5 faces	
6 corners	
9 edges	

Challenge

1 Draw real-life objects that are the same shape as these 3D objects.

3D object	Real-life object
sphere	
cone	
cylinder	
cube	
rectangular prism	

2

a Sort these 3D objects into groups. Give each group a heading. Then, under the correct heading, draw each object **or** write its name.

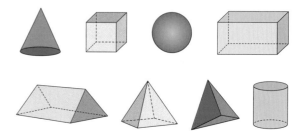

b How did you sort the objects? _____

_____

Oxford University Press

1 Guess which 3D object I am.

a I have _____ corners. I have five faces.

What am I? _____

b Write a "guess which 3D object I am" question for someone to solve.

What am I? _____

2 Draw some 3D objects. Write which 2D shapes make up each 3D object.

Did you know that a cube is made up of six squares?

3D object	2D shapes that make up the 3D object

Practice

1 Look at the picture and answer the questions.

a What is on the rug? _____

b What is under the bed? _____

c The photo frame is _____ the shelves.

d The rug is _____ the bed.

e The train is _____ the shelves.

2 Follow these directions to create a picture.

a Draw a sun in the top right-hand corner.

b Draw two flowers in the bottom middle.

c Draw a person next to the flowers.

d Draw a pond in the bottom left corner.

e Draw three ducks on the pond.

f Draw a butterfly above the person.

Oxford University Press

Challenge

1 Use the picture to describe the position of the items.

Item	Position description
shelves	
teddy bear	
clock	
window	

2 Add some items to the picture.

a Draw a table above the rug.

b Add a vase next to the train.

c Put a pillow on top of the bed.

d Draw a lamp on the table.

1 Look at the classroom map and answer the questions.

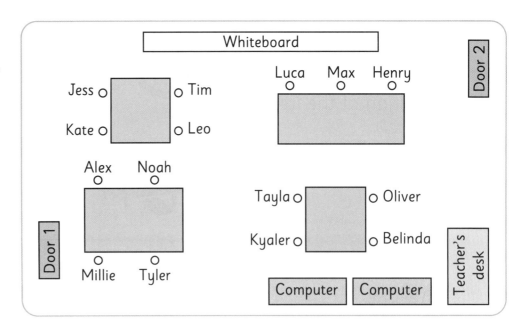

a Who sits furthest from door 1? _____

b Who sits closest to the teacher's desk? _____

c Who sits next to Leo? _____

d Write a question of your own for someone else to answer.

_____

2 Look at the picture below.

What language will you use? You might like to use "next to", "under" or "above".

Write the step-by-step instructions you would tell someone, so they could draw the same picture.

_____

_____

_____

Practice

1 Circle the objects below that have been turned anticlockwise.

2 Look at the map. Work out how many grid spaces from one place to another. For example, to get from school to the shop, go 2 spaces up and 2 spaces right.

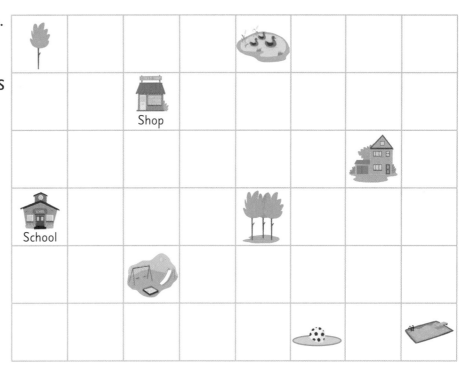

How many spaces from:

a   the tree to the pond? _____

b   the house to the swimming pool? _____

c   the playground to the sports oval? _____

d   the forest to the shop? _____

1 Use the map and follow the directions. Where do you finish?

Directions:

a Start at the tree.

b Go right 3 spaces.

c Go down 2 spaces.

d Go right 2 spaces.

e Go down 2 spaces.

f Go left 3 spaces.

g Where did

you finish? _____

2

a Using the map above, write your own set of directions for someone else to follow.

_____

_____

_____

b Where did they finish? Were your directions correct?

_____

_____

_____

1 Draw a map on the grid below. You should be able to use your map to follow these directions.

a Start at Leo's house in the bottom left corner and go 4 spaces right.

b Go up 3 spaces to Zoe's house.

c Go left 1 space and then up 2 spaces to Noah's house.

d Go right 4 spaces to Jack's house.

e Go down 3 spaces and left 1 space to Claire's house.

2 Use your map to write your own set of directions that will get you from one point on the map to another. How tricky can you make it?

Practice

1 On the graph, show how many people have each hair colour.

Number of people

15			
14			
13			
12			
11			
10			
9			
8			
7			
6			
5			
4			
3			
2			
1			

Blonde   Red   Black   Brown

Hair colour

2 The graph shows how many people like each sport. Use the graph to complete the table.

Soccer Basketball Rugby Netball Football Baseball

Sport	How many?

Oxford University Press

Challenge

1 Class 1 was asked what sports they play outside school. These are their answers.
Create a table to put this data into.

You may want to use tally marks!

Aidan	Basketball	Lucy	Netball
Josh	Football	Daniel	Basketball
Ava	Dancing	Holly	Dancing
Mo	Football	Sam	Karate
Charlie	Basketball	Chase	Football
Liam	Basketball	Ash	Dancing
Mya	Netball	Luca	Dancing
Laura	Football	Lily	Football
Logan	Football	Jake	Karate
Ella	Dancing	Jayda	Netball
Riley	Basketball	Harper	Football
Nico	Football	Jess	Netball

2 Use the table to draw a graph.

1 Now it's your turn.

a Write a yes/no question you could ask people.

b Write a question with four possible answers you could ask people.

2 Choose one of your questions from question 1 and collect the data.

a Your question _____

b Record people's answers. Tally how many people give each answer.

Answers	Tally marks

Great work! Now you can create a graph from the data you have collected.

Recording and interpreting data

Practice

1 Use tally marks to record the colours.

	Yellow	Red	Blue
Total			

2 Answer the questions about the graph.

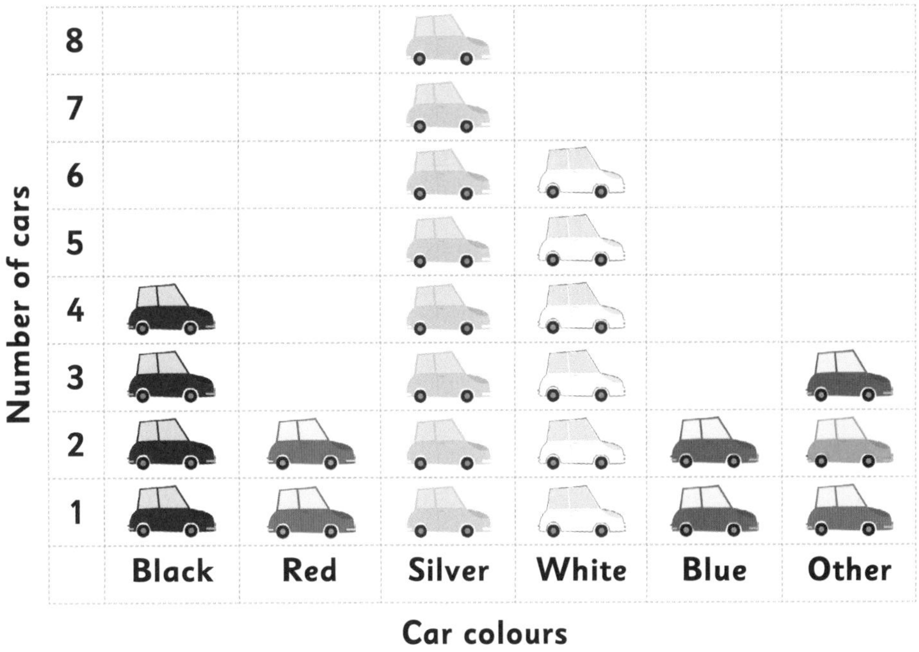

a Which is the most popular colour car? ____________________

b Which two colours have the same number of cars? ____________________

c How many cars were there in total? ____________________

d What might go in the "other" column? ____________________

1

a Choose a way to sort the shapes. Write the categories at the top of the tally chart.

b Use tally marks to record the number for each category. Write in the totals.

	Categories			
Tally				
Total				

2 Here is a completed graph.

a What do you think this graph is about?

_____

_____

b What information can you get from this graph?

_____

_____

_____

8				
7				
6			🌲	
5			🌲	🌸
4		🍂	🌲	🌸
3		🍂	🌲	🌸
2		🍂	🌲	🌸
1		🍂	🌲	🌸
	Summer	Autumn	Winter	Spring

c What title would you give this graph? _____

_____

Oxford University Press

1 Use this information to draw a graph.

- The most popular colour is blue with seven people.

- The least popular is brown.

- Two more people liked red than green.

- Pink and purple had the same with two less than blue.

- Three people liked yellow.

2 Below are Rahul's answers to some questions about the graph. Can you work out what the questions might have been?

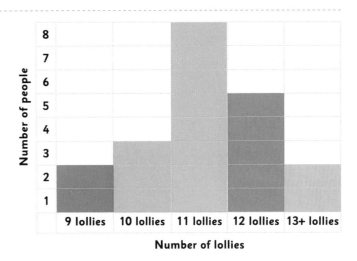

Answer	Question
8 people	
9 lollies	
9 and 13+ lollies	
7 people	
20 people	

Practice

1 Write in the boxes whether these **will** happen, **won't** happen or **might** happen.

A cat will read
the newspaper.

It will get dark.

It will be a
sunny day.

You will make
a new friend.

2

a Draw or write three things that **will** happen today.

b Draw or write three things that **won't** happen today.

Oxford University Press

1 For each bag, what is the chance of pulling out a blue marble? Write **will**, **won't** or **might**.

Bag of marbles	Chance

2 Draw groups of marbles to match the descriptions.

Description	Marbles
You won't pick a yellow marble.	
You will pick a blue marble.	
You might pick a green marble.	
Red has the most chance of being picked.	

Mastery

1

a Flip a coin six times. Predict what the coin will land on (heads or tails) before you flip it each time. Record your answers below.

Prediction	Actual result

b How many times were you right? _____

c How many times were you wrong? _____

d Describe the chance of getting it right. _____

2 Roll a dice 20 times. Before you start, predict how many times you think you will roll each number. Record the results as a tally.

	Number on the dice					
	1	2	3	4	5	6
Prediction						
Result						

How close were your predictions to your actual results?

Try this again! Are your results the same the second time around?

_____

_____

_____

ANSWERS

UNIT 1: Topic 1

Practice

1 a 41 **42** 43 **44** 45 46 **47** **48** **49** 50 51

b 95 96 **97** **98** 99 **100** 101 **102** 103 104 **105**

c 63 **62** **61** **60** 59 58 **57** **56** **55** 54 53

d **99** 98 97 **96** 95 94 **93** **92** 91 **90** 89

2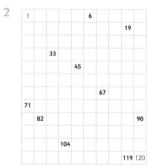

	1				6				
								19	
	33								
		45							
				67					
71									
	82							90	
		104							
							119	120	

Challenge

1 a 24   b 19   c 12
   d 21   e 7

2 Teacher to check. The difference should be 14. For example, 16 and 2, 24 and 10, 45 and 31.

Mastery

1 Teacher to check. Must be 6 numbers that add to 30, e.g. 4 + 5 + 8 + 4 + 3 + 6

2 a 20, 24, 42, 46, 64, 68, 86
   b 25, 47, 63, 69, 85

UNIT 1: Topic 2

Practice

1 seventeen, forty-one, sixty-six, one hundred and three

2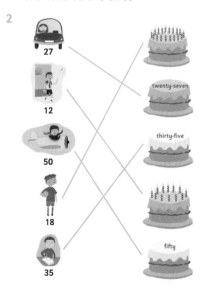

Challenge

1 Teacher to check.
2 Teacher to check.

3

Number line	Number (numerals)	Number (words)
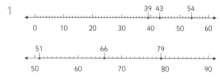 18 19 20 ✕ 22 23 24	21	twenty-one
106 107 108 109 ✕ 111 112	110	one hundred and ten
27 ✕ 25 24 23 22 21	26	twenty-six
43 42 ✕ 40 39 38 37	41	forty-one
89 ✕ 91 92 93 94 95	90	ninety

Mastery

1 Teacher to check.

2 16 sixteen   52 fifty-two
   25 twenty-five   61 sixty-one
   34 thirty-four   70 seventy
   43 forty-three

UNIT 1: Topic 3

Practice

1 39 43 54
0 10 20 30 40 50 60

51 66 79
50 60 70 80 90

2 Teacher to check.

Challenge

1 a–f Teacher to check.

2 a 27, 57, 24, 54, 52, 25, 42, 72, 45, 74, 47, 75
   b Largest to smallest: 75, 74, 72, 57, 54, 52, 47, 45, 42, 27, 25, 24

Mastery

1 a Yes, the scores will change the ladder.
   b

Team	Scores
Blue	34
Red	29
White	28
Green	27
Black	23
Pink	22
Brown	20

2 Teacher to check.

3 Teacher to check.

UNIT 1: Topic 4

Practice

1 Teacher to check.

2 a 20   b 19

3 a 16 + 4 = 20

0 1 2 3 4 5 6 7 8 9 10 11 12 13 14 15 16 17 18 19 20

   b 9 + 8 = 17

0 1 2 3 4 5 6 7 8 9 10 11 12 13 14 15 16 17 18 19 20

Challenge

1 Teacher to check.
2 Teacher to check.
3 Teacher to check.

Mastery

1 Teacher to check. Julie must have 11 points more than Maria, e.g. Julie 24 and Maria 13.

2 Teacher to check. Paul's birthday must be 14 days after Bill's, e.g. Bill on 2 June and Paul on 16 June.

UNIT 1: Topic 5

Practice

1 a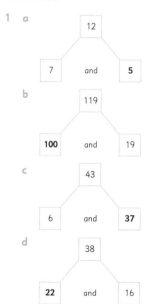

12
7 and 5

b
119
100 and 19

c
43
6 and **37**

d
38
22 and 16

ANSWERS

2

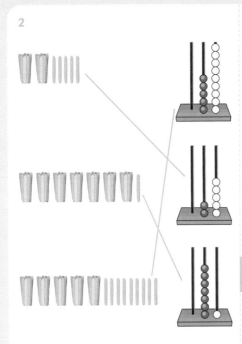

Challenge

1 Teacher to check. The partitioned numbers must add up to the number in the "Number" column, e.g. 13 = 10 and 3, 5 and 8, or 6 and 7.

2 Teacher to check. The amount she gave away must be half the number she started with.

Mastery

1 Teacher to check. Examples include 100 and 10, 50 and 60, 90 and 20.

2 Teacher to check. The difference must be 9, e.g. 24 and 15, 30 and 21, 10 and 1.

UNIT 1: Topic 6

Practice

1 a 11

   b 16 − 7 = 9, 14 − 9 = 5, 19 − 11 = 8

2

Squares uncovered	Peter's score (squares covered)
3	17
6	14
11	9
14	6
9	11

Challenge

1 Teacher to check.

2 Teacher to check. Answers will vary depending on the numbers chosen by students. Look for whether students can accurately answer using the 0–99 chart by taking away 10s then 1s.

Mastery

1 Teacher to check. For example, 32 − 8 = 24

2 Teacher to check. For example, 50 − 9 = 41, 45 − 9 = 36, 30 − 9 = 21

UNIT 1: Topic 7

Practice

1 a 8   b 6   c 8

   d 12   e 16   f 16

2 Teacher to check. Students choose two numbers and use the number line to add or subtract to accurately find the difference.

Challenge

1 Teacher to check. The difference needs to be 7, e.g. 23 and 16.

2 Teacher to check.

Mastery

1 Teacher to check. The red team's score needs to be more than double the blue team's score.

2 Teacher to check. For example, 38 is 7 more candles.

UNIT 1: Topic 8

Practice

1 a 25   b 40   c 17

2 Eyes: 24, Eyes: 46, Eyes: 38

Challenge

1 a 0, **5**, **10**, **15**, 20, **25**, 30

   b **10**, 15, **20**, **25**, 30, **35**, **40**

   c 18, **20**, **22**, **24**, **26**, 28, **30**

   d **55**, **50**, 45, **40**, **35**, 30, **25**

   e **60**, 50, **40**, **30**, **20**, **10**, 0

2 a Teacher to check. Skip count by 2s.

   b Teacher to check. Skip count by 5s.

   c Teacher to check. Skip count by 10s.

3 a Teacher to check. Skip count backwards by 2s.

   b Teacher to check. Skip count backwards by 5s.

Mastery

1 Teacher to check. For example, skip count by 5s: 30, 35, 40, 45, 50.

2 a Teacher to check. Skip count by 2s.

   b Teacher to check. Skip count by 10s.

UNIT 1: Topic 9

Practice

1 a 5   b 4   c 3   d 4

2 Teacher to check. Equal shares of 4 each, e.g. 3 friends = 4 + 4 + 4 = 12.

Challenge

1 a 6   b 4   c 3   d 2

2 a Yes, they get 3 each. Teacher to check working out.

   b No. Teacher to check working out.

Mastery

1 Teacher to check: 20, 10, 5, 4, 2, 1.

2 Teacher to check. They must get 3 each, e.g. 5 friends = 3 + 3 + 3 + 3 + 3 = 15.

UNIT 2: Topic 1

Practice

1

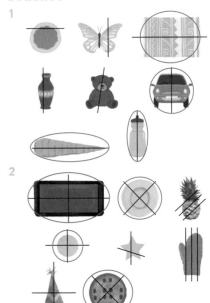

2

Oxford University Press

Challenge

1 Possible answers:

2 a Possible answers:

b Possible answers:

c Possible answers:

Mastery

1 Teacher to check.

2 Teacher to check.

UNIT 2: Topic 2

Practice

1 a Half of each group should be circled, e.g.

b

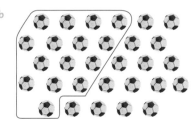

---

2 One quarter of each group should be circled, e.g.

a

b

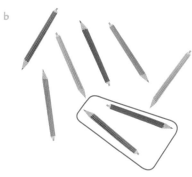

3 a 2
  b 6

Challenge

1 Teacher to check. Must be half of each type of cookie.

2 Teacher to check. One quarter of the marbles must be blue.

Mastery

1

Question	Working out
There are 24 students in Sienna's class. Half are boys. How many boys are there?	12 (Teacher to check working out.)
Dan has 20 people in his class. $\frac{1}{4}$ of them are away today. How many are away?	5 (Teacher to check working out.)
There are 30 people watching rugby. Half are wearing jumpers. How many people are wearing jumpers?	15 (Teacher to check working out.)
Jess has 16 people in her class. $\frac{1}{4}$ are away today. How many of her class are at school?	12 (Teacher to check working out.)

---

2 Teacher to check. For example:

Total counters	Red counters	Green counters	Yellow counters
20	10	5	5
16	8	4	4

UNIT 3: Topic 1

Practice

1 a Teacher to check. Students should draw $2, $1, 50c, 20c, 10c, 5c.
  b 2   c 4   d 50c
  e 5c   f $2   g 5c

2

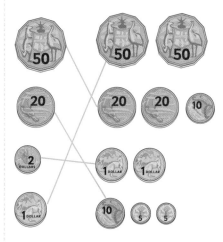

Challenge

1 Teacher to check. For example, 50c + 50c, 50c + 20c + 20c + 10c, 20c + 20c + 20c + 20c + 10c + 10c.

2 a

b

c

d

ANSWERS

Mastery

1　Teacher to check.
2　Teacher to check. For example, $2, 50c, 10c, 5c or $1, $1, 20c, 20c, 20c, 5c. Students should identify how many coins they have.

UNIT 3: Topic 2

Practice

1　a　

　b

2　a　20c and 50c
　b　$1, 10c, $2, 5c
　c　$1, $2
　d　20c, 10c, 5c
　e　$10c \times 2 = 20c$,　$5c \times 2 = 10c$, $1 \times 2 = 2,　$50c \times 2 = 1

Challenge

1　One possible solution:

Item	Draw the coins (most value to least value)
$1.50	
$3.00	
$2.25	
$3.95	
$5.10	

2

Coins	Order	Total
		85c
		$3.60
		$2.00
		$1.35
		$3.75

Mastery

1　a　$10c \times 2 = 20c$, $5c \times 2 = 10c$, $1 \times 2 = 2, $50c \times 2 = 1
　b　Teacher to check. For example, $10c \times 5 = 50c$, $10c \times 10 = 1, $50c \times 4 = 2
2　Teacher to check. Two of the three coins must be the same.

UNIT 4: Topic 1

Practice

1　a
　b
　c　M W A M W A M W A M W
　d　P P R P P R P P R P P R

2

Circle the error in each pattern	What should it be?	Rule for the pattern
↑↑↓↑↑↓↑	↑	2 arrows up and 1 down
●●●●●●●●	●	red circle, blue circle
X Y Z X Y Z X X Z	Y	X, Y, Z
O U U O O U O O U	O	O, O, U
I I J I I I J J I I J J	J	I, I, J, J

Challenge

1　Teacher to check.
2　a

　b

　c

Mastery

1　Teacher to check.
2　Teacher to check.

UNIT 4: Topic 2

Practice

1　a　36, 38, 40, 42, 44, **46**, **48**, **50**, **52**, **54**
　b　55, 60, 65, 70, 75, **80**, **85**, **90**, **95**, **100**
　c　120, 110, 100, 90, 80, **70**, **60**, **50**, **40**, **30**
　d　80, 78, 76, 74, 72, **70**, **68**, **66**, **64**, **62**
　e　Pattern a: goes up in 2s, Pattern b: goes up in 5s, Pattern c: goes down in 10s, Pattern d: goes down in 2s.

2　a　60, 70, **80**, 90, **100**, **110**, 120, 130, **140**
　b　110, 105, 100, **95**, **90**, 85, **80**, **75**, 70
　c　92, **94**, 96, 98, **100**, **102**, **104**, 106, 108
　d　190, **180**, **170**, 160, **150**, **140**, 130, 120
　e　Pattern a: goes up in 10s, Pattern b: goes down in 5s, Pattern c: goes up in 2s, Pattern d: goes down in 10s.

Challenge

1　a–c　Teacher to check.
2　a　Teacher to check. Pattern must be counting by 10s, not starting from 10.
　b　Teacher to check. Pattern must be counting by 5s, not starting from 5.

Mastery

1　a–b　Teacher to check.
2　Teacher to check.

UNIT 5: Topic 1

Practice

1

2　a　18 tiles　b　16 tiles
　c　8 tiles　d　12 tiles

Oxford University Press

Challenge

1 Teacher to check.

2 Teacher to check.

Mastery

1 Teacher to check.

2 Teacher to check. For example:

UNIT 5: Topic 2

Practice

1 a 5 cubes b 8 cubes

  c 9 cubes d 20 cubes

  e Smallest to largest: a, b, c, d
    (or 5, 8, 9, 20).

2

Challenge

1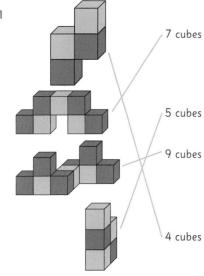

7 cubes

5 cubes

9 cubes

4 cubes

2 a Teacher to check.

  b Teacher to check.

Mastery

1 Teacher to check. Objects should each have a volume of 8 cubes, e.g.

2 a–b Teacher to check.

UNIT 5: Topic 3

Practice

1 a   b

  c   d

2 Teacher to check.

Challenge

1 a Teacher to check.

  b Teacher to check. Do children accurately arrange their items from lightest to heaviest?

2 a–c Teacher to check.

Mastery

1 a–d Teacher to check.

2 a 6 cubes b 8 cubes

  c 7 cubes d 13 cubes

UNIT 5: Topic 4

Practice

1

 to

 to

 to

 to

2

Digital time	The time in words	Analog time
11:30	eleven-thirty	
6:00	six o'clock	
9:00	nine o'clock	
5:30	five-thirty	
1:30	half past one	

Challenge

1 a 2:30

  b 11:00

  c 9:30

  d 4:30

  e 2:00

  f 8:30

2 a b c

  d e f

Mastery

1 a–b Teacher to check.

2 Teacher to check. For example, 6:00 or anything half past the hour.

UNIT 5: Topic 5

Practice

1 a, d, c, b

ANSWERS

2

Event	Time measured in
Your next birthday	Teacher to check. Days, weeks or months.
You finish primary school	Teacher to check. Likely answer is years.
The weekend	Teacher to check. Hours or days.
Winter	Teacher to check. Days, weeks or months.

Challenge

1   Weekend by the sea = days
    The school holidays = weeks
    A school day = hours
    Play on a seesaw = minutes

2   Teacher to check.

Mastery

1   Teacher to check.

2   a–b   Teacher to check.

UNIT 6: Topic 1

Practice

1   a–c   Teacher to check. For example, group shapes with more or fewer than 3 sides.

2   a–b   Teacher to check.

Challenge

1   a   4 corners. I am a rectangle.

    b   Teacher to check.

2   Teacher to check. Look for descriptions such as the number of corners or sides, and whether sides are parallel.

Mastery

1   Teacher to check.

2   Teacher to check.

UNIT 6: Topic 2

Practice

1

cube   rectangular prism   cone   sphere   cylinder

2   Teacher to check. Examples of possible answers are below. Check drawings.

Description	Draw and name the object
6 faces	cube
0 edges	sphere
5 faces	pyramid
6 corners	triangular prism
9 edges	triangular prism

Challenge

1   Teacher to check. Examples of possible answers are:

3D object	Real-life object
sphere	ball, marble
cone	party hat, ice cream cone
cylinder	drink can, candle
cube	dice, ice cube
rectangular prism	tissue box, brick

2   a–b   Teacher to check. For example:

5 or more faces	Fewer than 5 faces
cube	cone
rectangular prism	sphere
triangular prism	cylinder
square-based pyramid	
triangular-based pyramid	

Mastery

1   a   I have 5 corners. I am a square-based pyramid.

    b   Teacher to check.

2   Teacher to check, for example:

3D object	2D shapes that make up the 3D object
cube	6 squares
square-based pyramid	1 square, 4 triangles

UNIT 7: Topic 1

Practice

1   a   book   b   shoes

    c   above   d   next to

    e   on top of

2   a–f   Teacher to check.

Challenge

1   Answers may vary but could include:

Item	Position description
shelves	next to the rug
teddy bear	on top of the bed
clock	above the bed
window	next to the clock

2   a–d   Teacher to check.

Mastery

1   a   Henry   b   Belinda

    c   Tim   d   Teacher to check.

2   Teacher to check.

UNIT 7: Topic 2

Practice

1

2   a   4 spaces right

    b   3 spaces down and 1 right (or 1 right, 3 down)

    c   1 space down and 3 right (or 3 right, 1 down)

    d   2 spaces left and 2 up (or 2 up, 2 left)

Challenge

1 At the playground.

2 a–b Teacher to check.

Mastery

1 a–e Teacher to check.

2 Teacher to check.

UNIT 8: Topic 1

Practice

1

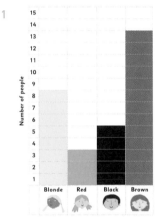

2

Sport	How many?
soccer	3
basketball	7
rugby	2
netball	4
football	9
baseball	1

Challenge

1

Sport	No. of people
basketball	5
netball	4
karate	2
football	8
dancing	5

2 Teacher to check. Graph may look like this:

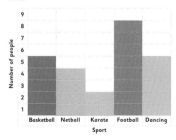

Mastery

1 a–b Teacher to check. Look for whether students can ask questions that meet the given criteria.

2 a–b Teacher to check.

UNIT 8: Topic 2

Practice

1

	Yellow	Red	Blue
	⦀⦀ ⦀⦀ ⦀⦀⦀⦀	⦀⦀ ⦀⦀	⦀⦀ ⦀⦀⦀
Total	14	10	8

2 a silver

  b red and blue

  c 25

  d Car colours that don't have their own column on the graph, like green or yellow cars.

Challenge

1 a–b Teacher to check. Categories could include shapes, colours or sizes.

2 a Teacher to check. For example, it's about which is the most/least popular season.

  b Teacher to check. For example, that summer got the most with 8 and autumn the least with 4.

  c Favourite seasons.

Mastery

1 Teacher to check. One possible answer is:

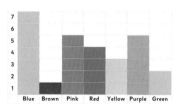

2 Teacher to check. One possible answer is:

Answer	Question
8 people	How many people had 11 lollies?
9 lollies	What is the least amount anyone had?
9 and 13+ lollies	Which 2 amounts had the same?
7 people	How many people had 12 lollies or more?
20 people	How many people completed this task?

UNIT 9: Topic 1

Practice

1 won't, will, might, might

2 a–b Teacher to check.

Challenge

1

Bag of marbles	Chance
	might
	will
	won't
	might

2 Teacher to check. One possible answer is:

Description	Marbles
You won't pick a yellow marble.	For example, 4 blue marbles.
You will pick a blue marble.	For example, 4 blue marbles.
You might pick a green marble.	For example, 1 green, 2 blue, 2 yellow marbles.
Red has the most chance of being picked.	For example, 4 red, 2 yellow marbles.

Mastery

1 a–d Teacher to check.

2 Teacher to check.